MONETIZE YOUR CONTENT

HOW TO CREATE AND PRODUCE YOUR OWN
SHOW

BY: LEE HARRIS

D1412386

This book is dedicated to my grandparents, my parents, Chloe, Naomi, my family, Adrienne and those who believed in me during my journey. Thank you.

Table of Contents

CHAPTER 0 PREFACE

THE INTRO TO THE INTRO...

First off, thank you for reading this book. Hopefully you bought it, borrowed it or loaned it from a friend. If so, I'm truly thankful. If you loaned it I'm telling you now, go out and buy your own copy, it's worth it! If nothing else; you'll get a thank you every day. First sentence! Now, how many times does someone thank you for your contribution on this earth. Well, I'm here to say thank you, bless your creative spirit. The goal here is to motivate, inspire and spread positivity.

Now, there are probably three reasons why you picked up this book...

1- You have a great show idea. Now, you want to know how you can create and produce it on your own. Kudos to you!
2-You had an idea. Tried to make a show and didn't make it to the end. You want some professional tips and a clear strategy on how to get from start to finish. You're in the right place!

3-You heard about people making money on YouTube, Netflix, iTunes and Amazon. Now you want in on the action. Dreamers must turn dreams into goals! Well this book is for you too!

WHAT IF I JUST WANT TO STIMULATE MY CREATIVITY AND LEARN HOW TO FINISH PROJECTS?

You are absolutely correct. This book is also for people who want to stimulate their creativity. People who want to learn how to start projects and finish them. Oh, and everyone else who bought the book too!

Thank you. Best wishes enjoy the journey!

Lee Harris

CHAPTER 1: INTRO

TIME FOR THE SHOW

This book is designed to empower content creators. My mission is to motivate, inspire and educate creative people. We all have ideas that pop into our heads. Once an idea is planted inside of your head it's your job to bring that idea into reality.

My goal is to prepare you for the journey and to arm you with tools that will make the journey easier. Some of you have ideas but are not quite sure how to execute your plan. This book is designed to guide you through the major stages of production which are pre-production, production (shoot days) and post-production (editing & final delivery). I will also discuss video sharing strategies and monetization options for your project.

In other words, we'll help you create your show idea, write a treatment, pitch the concept, hire a crew, select equipment, shoot, film and edit the show. You are content creators. You are individuals with a story to tell. I am presenting these ideas to both the novice and the professional because everyone needs a little support on their journey.

OK, HOW ARE YOU QUALIFIED TO GUIDE US ON THIS JOURNEY?

Great question. I started out at MTV as an intern, while attending Rutgers University. After graduating I was hired as a writer and quickly rose through the ranks of the network due to a strong work ethic and unrelenting passion for production, music and pop-culture. While there I launched the career of a talented young singer named Tyrese Gibson who became host of MTV Jams. We are still close friends today. I also worked on MTV Beach House, Sports & Music Festival, TRL, New Year's Eve 2 Large and Fashionably Loud. It was an amazing experience. When first starting out as a producer I was fortunate enough to work around some great television minds. People like Tony DiSanto who created TRL, Laguna Beach and The Hills. And, Dave Sirulnick who ran the MTV News and documentaries department, which was home to series like 'Made', 'Tru Life', 'Week in Rock' and 'Diary'. He was also executive producer of the 'MTV Video Music Awards' and 'TRL'. I was young, but I studied, listened and learned a lot.

In the fall of 2000, I took a leap of faith and left MTV to co-create BET's long-running flagship show, 106 & Park. A few years after launching 106 & Park, I was on the executive staff that helped re-launch MTV2 with hit shows like Guy Code, Girl Code and the Sucker Free Awards.

Over the past 15 years I've been able to produce, write and direct Award shows, tv shows, documentaries, talk-shows, comedies, music specials and sporting events. My work includes creating a diverse array of programming for television networks such as MTV, BET, OWN, MTV2, ABC, TV One, Revolt, VH1 and film companies like Blumhouse Productions. I have experience writing and producing shows for top celebrities such as Beyoncé, Dave Chapelle, Vin Diesel, Channing Tatum, Harrison Ford, Denzel Washington, Jay-Z, Queen Latifah, Will Smith, Tom Cruise, Seth Rogen,

Will Ferrell, Sam Jackson, Bruce Willis, Traci Ellis Ross, Lauryn Hill, Jonah Hill, Kevin Hart, Kanye West, Pharrell, Common, Nicki Minaj, Anthony Anderson, Wendy Williams, Lil Wayne, Shaquille O' Neal, Jamie Foxx, Kobe Bryant, Ice Cube, Issa Rae, Floyd Mayweather Jr., Ludacris, Snoop Dogg, Sean 'Diddy' Combs, Outkast, Sway, Carson Daly, Bill Bellamy, Charlamagne, Wyclef, SZA, and RZA, just to name a few. I truly appreciate each experience and grow with every interaction. Each person works differently, and it's my job to get the best performance from them during a production. You must make them comfortable when appearing on camera, whether they are promoting a project, playing a character or reading lines from a script. As a youth growing up in Staten Island, NY, who would have imagined that I would get to work with some of my favorite celebrities. I'm thankful and would like to share what I've learned along the way.

Hopefully that answers your question about my qualifications. Now that we've discussed the past let's jump back into the present. This is your future.

So, you have a great idea. What's next? What's your plan?

Are you thinking about becoming a YouTube sensation?

Perhaps you want to get your show on a video streaming platform like Netflix or Amazon?

Or, maybe you want to sell your project to a major network?

This book will show you how to do it. I will guide you through the entire process. The goal is to help you create the show you dreamed about making. Whether you are a novice or an experienced professional, this book will help you understand the entire production work flow. You can view it as an advanced producer course. If you have experience working in the television or film industry but always wanted to produce your own show, this book will make you look like a veteran. Always remember, the more you practice, the better you'll get. Feel free to read a chapter multiple times or bring the book on the production set with you. This book is your secret weapon, when in doubt, pull my book out. It's time to put the power in the hands of creative people. I know that you can do it and I believe in you. Let's get going!

CHAPTER 2 YOUR IDEA – WHAT IS IT?

It's time for a few important questions. Let's turn your idea into money making content!

What exactly is your show or project about?

What kind of show is it?

When answering these questions think about which genre you're aiming for. There are so many different genres. It's like when you're on Netflix or Amazon looking for tv shows, films or music. They have categories that represent different genres. Each one makes it easier for a fan of the genre to make a proper selection. The same is true for you and your project.

It's important that you figure out which genre your show fits into. Is the genre comedy, drama, technology, sports, news, action, cooking, adventure, animation, horror, documentary, biography, crime, fantasy, historical, suspense, thriller, a western, a performance special or a pod cast? Make sure you know which genre your show is going for.

BUT WAIT, WHAT IF I DON'T HAVE A NEW IDEA?

If you don't have a new idea, that's fine. There are options, you can create something entirely new or put a new twist on an old show. Some people say that "everything has already been done" and there's no original idea. Well, that's pretty limiting but if that's what you believe to be true, then every idea is a twist on an old idea. So, feel free to put a new twist on something. Especially if as a creator, you're not happy with the version that has already been made. Just make it your own and bring something new to the table. Your new twist might make it an original creation for your audience.

WELL, CAN YOU GIVE ME SOME EXAMPLES?

No need to stress, if you need some examples of what I'm talking about, I've listed some below. Here are some examples:

The hit show *MTV Cribs* was a fun re-invention of the classic show 'Lifestyles of the Rich and Famous' created by Alfred M "Al" Mansini. So is the Fabulous Life Of...

Shonda Rhimes, the supreme Queen of Thursday night drama, admits that her original pitch of Grey's Anatomy was a new twist on the hit show Sex In The City. She even pitched it as, Sex In The Surgery. Some say Grey Anatomy is a multi-cultural twist on the hit show E.R. Either way Shonda makes amazing shows and is extremely talented.

'In Living Color' was a new twist on Saturday Night live. Mad TV was a new twist on 'In Living Color'.

When we created a show called 106 & Park for B.E.T, it was a new twist on two of my favorite shows, the 'Arsenio Hall show' and 'TRL'. That's why I was adamant about having a couch, Arsenio Hall had one! I wanted that late-night talk-show vibe. When I created a segment called Freestyle Friday, it was a twist on the hip-hop rap cyphers that I experienced as a youth. When experiencing these cyphers, I loved the witty word play and self-expression found in hip hop culture.

Can you think of some more examples? I'm sure you can. Feel free to send them to me if you do. You get the point, some of the best shows are based on an idea that already existed. The audience just needed a new twist. A new perspective.

WHAT IF THAT DOESN'T WORK FOR ME? ANY OTHER OPTIONS?

Think about any holes in the market, any places where you could create content to fit that specific niche and then do it the best you can. Remember, there was a time where sports fans needed more sports content and ESPN filled that niche. Comedy Central created a network focused on the niche of people who wanted more comedic content. The Food Network created shows for people who love to cook, eat and learn more about food.

Cable originally differentiated itself from network television by featuring channels that served a specific niche. Now we have the internet and that's what websites do. Each one has a purpose, a specific audience and a niche that it serves.

It's all about filling that hole and creating a show that fits and satisfies your niche. If you create the content and fill that void, the audience will reward you with viewership, loyalty and support. Don't hesitate, they're waiting for it!

Exercise

List three (3) of your favorite shows. Then write down which genre they fit into.

Now, time for your show. Think about which genre it fits into and write it down.

CHAPTER 3: WHO ARE YOU MAKING THIS FOR? BUILDING YOUR AUDIENCE

You have a show idea. Can you answer the following questions?

Who is it for? Why do they want it? Why do they need it?

When you're thinking about your project it's great to consider who you're making it for. Who your core audience will be. You're probably thinking, my audience is everyone! I'm going to have a hit show with billions of viewers! And, that is possible, we are not here to squash dreams. But let me accompany that with a tactic that will help you achieve your grand plan, the key to getting those viewers is to start small with a core audience.

Quick story: When I was supervising producer at MTV2 there was a show being developed by a friend of mine, a talented showrunner and producer named Ryan Ling. The show was called Guy Code. Before he went into production we had some great conversations about the show and what he was aiming for. His goal was to give guys like him hilarious advice about dating and life.

Ryan Ling would stop by my office to discuss the direction of the show and who he wanted to bring in. Once the project really started making progress MTV gave him a small team plus office space. I would stop by his office and he would tell me about the specific kinds of comedians he wanted, different re-enactments and topics they wanted to cover. The topics. He had an incredible board of topics and we would crack up discussing which ones would work best on the show.

Well, as you know, his show worked, and it was the most successful show in MTV2 history. As a matter of fact, it was so popular, it launched a spin-off series called Girl Code. Ryan won because he had a great idea, he knew his niche, locked into his creative and executed at a high level. The name of the show is great too. A perfect storm. A powerful example of how understanding your audience and your niche can work out extremely well. Also proof of why great titles kick booty!

Marketing guru Seth Godin developed a theory called the 'Idea Virus'. His theory is that ideas spread most effectively from customer to customer. Prior to this, marketing was based on an older model, the commercial or advertisement-based model. That theory stated that ideas spread best from business to customer. That is no longer true. Businesses cannot rely on simply paying for a commercial on TV or radio and having that commercial bring in millions of customers. In this digital age, word of mouth on social media can make or break a business. Customers sharing positive experiences with other customers is the way products really explode! That is why Yelp! Is huge.

The same is true for the project you're creating. Let's call it a Show Virus. If you get a loyal group of fans who appreciate your work, they will spread the word to the masses. The best way to get to create a 'show virus' is to start small. Don't try to please everyone, because that's impossible. Pleasing everyone is like making baby food. Only babies eat baby food. Try eating baby food. No way. That's because our taste buds require flavor, something special, maybe even something spicy. That's not baby food. Making something for everyone is like eating baby food, it's so bland it has no taste or flavor at all. Sorry Gerber!

Create a small niche group and work on making them happy with your content. Once you gain their trust they will begin to work for you, spreading your content via social media, e-mail and word of mouth.

When you create a community and give people a chance to be part of something bigger than themselves, they'll welcome the opportunity to share it. Most of the popular YouTube celebrities built up their audiences organically. With talent, focus, hard work and lots of videos. It's impressive to see the work ethic of those YouTube creators. They're hungry, passionate and will do whatever needs to be done to make the show work.

When cable television dominated the industry, it was popular because it had networks that appealed to specific audiences. A history channel for history buffs, a weather channel for people who love weather updates and youth channels devoted strictly to children's programming. These networks aren't successful because they chose to please everyone. The key to success was that they found a target audience and they maximized their relationship with that target audience. The internet has taken that same theory to the next level. Now there are social networking sites, blogs and millions of websites geared towards individuals with extremely different tastes.

As human beings we enjoy being a part of a tribe or community. It's in our DNA. Whether its family ties, social networks, school groups, work cliques or professional organizations, we enjoy being part of a community. Events like Comicon, SXSW, concerts, auto shows, fashion shows, sporting events and film festivals are all about community. Your job as a content creator is to activate that community and make them want to support you.

DO YOU HAVE ANY TIPS FOR RESEARCHING AN AUDIENCE?

Let's take a deeper look at your audience research. There are two ways to gather more information about the group you would like to target; utilizing *demographics* and *psychographics*.

Demographic information: When dealing with demographic information for your audience you should consider the following factors: Age, gender, cultural background and ethnic background.

Psychographic information: This information analyzes personal attributes such as: Personality, Attitudes, Values, Interests/Hobbies, Lifestyle and behavior.

Most content creators have an idea of who they want their audience to be. Even when you're just making a video for your friends, you're creating content for a specific group. If you are creating the show for an audience like you, then I recommend that you answer the questions first. Do a test study on you, not as the show creator, but as a viewer. What would you want to see?

Once you can answer these questions, you'll have a better understanding of your target audience. It's important to keep these things in mind before you create your project. With this research at your fingertips it will be easier to make a show that resonates with your viewers. It's your duty as a content creator to make that happen.

EXERCISE – Take a moment to answer the following questions.

Who is your target audience?

What do you know about them?

Does your show fit with their needs?

How will the characters in your show appeal to your audience?

What will your audience get from your show?

CHAPTER 4: STUDY THE COMPETITION | DO YOUR RESEARCH

Be a great student and do your research. The best professionals in music, television and film do it and so should you!

You can't become one of the best unless you study the best. That list will be different for everyone. While studying film at NYU, I gravitated to films by Akira Kurosawa, Fellini, Godard, Martin Scorsese, James Cameron, Ridley Scott, Spike Lee, Hitchcock and Sergio Leone. Not only did I study their movies; I also read and watched any interview that was available.

One thing I noticed was how much they knew about directors and films that came before them. People like Spielberg, George Lucas, Christopher Nolan, Martin Scorsese, James Cameron, Spike Lee, JJ Abrams, Kathryn

Bigelow and Quentin Tarantino can tell you all about their favorite films, classic scenes and amazing performances.

The best in the industry don't just create films they are students – they enjoy studying others in their industry. The same thing is true in music. Top musicians are students too. Study your field. It will pay off!

As a writer or director in film school you are taught this simple strategy. Once you pick a style of film (whether it's romance, horror, action, a buddy flick or a comedy), you must watch films that are similar in the genre. This way you can take notes, see what worked and what didn't work. Studying films that are similar to yours helps you avoid the cliché. If you want to move the genre forward, you must study and know the history of that tradition.

People often talk about making a show, film or podcast but when asked about similar shows in the genre, they haven't watched any of them. They haven't studied the competition at all. How can you expect to compete or win if you don't know about the competition in your field?

Everyone at the top of their craft knows about their peers. Not only will it fuel your fire, it will spark ideas about what to do, what not to do and ways you can make improvements.

EXERCISE

Watch three (3) shows that are similar to your project. They don't have to be the same, just in the same genre. Take notes. After each show answer the following questions.

1. Who are your competitors targeting?

2. Can you define their target viewing audience?

3. Is the show like yours? If not, how is it different?

4. Is it the same audience you're going for? Why or why not?

5. What do you like about the show?

6. What would you improve?

7. How would you add your own twist?

8. Can you find a different niche? You may not want to go after the exact same market when you can find an original niche that's better suited for your success.

CHAPTER 5: WRITING THE IDEA. THE SHOW TREATMENT AND LOG LINE.

This chapter is all about creating a show treatment and a memorable logline. After the last chapters you should have a clearer idea about the show you want to make and who you want your audience to be. If not, please refer to the previous chapters.

Next step is creating a show treatment. A show treatment is a document that outlines the show. A three to five-page detailed plan of what your show is about and how you will produce it. It gives the reader a clear out line of crucial show elements including your show title, cast, setting, locations and show elements.

A treatment is extremely important if you're pitching your show to digital platforms, networks or distributors. Even if your plan is just to get some investors or sponsors, a show treatment is important.

WHAT IF I'M JUST UPLOADING MY SHOW TO YOUTUBE?

Never underestimate the power of YouTube. As many creatives know, YouTube is a powerful channel. Think of it as a digital television network and prepare for it the same way you would for any platform. A show treatment helps you organize your creative vision. The people that create for YouTube with a professional attitude and vision are the ones that make it the furthest. Plus, once you become popular on YouTube, you'll need to prepare for the next steps; merchandise, a book deal, a show on Netflix or Hulu. Who knows? Just look at people like Lilly Singh aka Superwoman, and Colleen Ballinger aka Miranda Sings.

It's always good to have your show treatment prepared. As the saying goes; stay ready, so you don't have to get ready. Sometimes there's no time to get ready. Enough of that, let's start working on your show treatment.

WHAT KIND OF SOFTWARE DO I NEED TO WRITE MY TREATMENT?

Good question. Here's a quick breakdown of software you can use to make your show treatment. These are my recommendations.

MS Word:

With Microsoft Word you can insert images, pictures, tables and write any copy that you'll need. In the past it was tricky using MS word because there were challenges adding images, but recent versions have helped alleviate those issues. As tempting as it is to just copy and paste pictures into the document (something I used to do all the time) my designer friends told me that the best way to do it is to 'insert' pictures. So please take their advice. Make sure that you insert your picture when using MS Word. Anyone with basic office software can get a deck done using MS Word.

MS PowerPoint

Microsoft PowerPoint is a program designed to make office presentations. Since it was created to make slide shows and presentations this program is a step up from MS Word (when it comes to making treatments). Personally, I still use MS word to write the initial text copy, then copy and paste that information into PowerPoint. It's easy to add, move and manipulate images in this program. Plus, it has a bunch of fun features, effects, templates and colors that will help enhance your creative. Each time you use it, you'll get better and more comfortable with it.

With practice you will get more comfortable, so don't get frustrated if there are a few learning curves. You'll have fun learning. Also, if you don't feel like you're super creative, you can start by using some pre-built templates. These will help guide your path and give you a nice foundation to work from.

Adobe Photoshop

This program is incredible at manipulating images and adding stylized text. Unless you are an experienced user and have the budget to afford the software I recommend using MS word or PowerPoint. That said, if you want to create a cover page or add special images Photoshop is one of the best programs to use.

CREATING A GREAT TITLE FOR YOUR SHOW

Now that we've discussed the best software to use the next big step is locking in a title for your show. This title is the one you will use on all documents and the one you will use when discussing the project. When coming up with a name for your show make sure that it's fun to say and easy to discuss. The show title should be witty; something catchy and memorable. Think of your favorite shows. It's about creating a show title that works and makes people more interested in the project.

Here are some examples from popular shows:

When you hear titles like The Walking Dead, Game of Thrones, Survivor, Boardwalk Empire, Entourage, The Big Bang Theory, Scandal, Empire, Hawaii Five O, Basketball Wives, Duck Dynasty, Shark Tank, The Bachelor, Guy Code and Undercover Boss, you get an idea of the show theme.

Once you know about the show, you really get an understanding of how clever the titles are. Plus, they give you a little insight as to what the show will be about. If you hear a show title like Walking Dead, you don't assume that it's a show like Sesame Street. Quite the opposite.

So, when you're thinking of your show title don't forget these tips: be clever, keep it fun, make it catchy and make it memorable. The first thing people will talk about when mentioning your show is the show title so make it count.

WRITING THE SHOW TREATMENT

Once you have a show title that inspires you, the next step is to write the treatment. Normally a treatment is 3-5 pages long but if there are additional images added, it can be a few pages longer. You want to make sure that your reader is captivated and engaged. The last thing you want is someone becoming uninterested in your project because the treatment was too long. You want to grab the reader's attention and create a need for them to want more.

SO REALLY, WHAT'S IN THIS SHOW TREATMENT?

Your show treatment should include an explanation of the show, the cast, setting details and important locations. You should describe the elements involved in your show. When dealing with the cast you should include a quick breakdown of what each person will do and what their role will be in the production.

The same is true for your set breakdowns. If your project is dependent on a specific set, provide a brief explanation of the set and the role it will play in your project.

BREAKING DOWN YOUR SHOW FORMAT. WHAT KIND OF SHOW IS IT?

Now this part is pretty straight forward. Think about the kind of show you're going to create. A very important aspect of every show is the genre or format it fits into. It's important to differentiate whether it's a competition show, game show, reality show, sports show, horror, sci-fi, podcast, action, crime show, comedy etc. You get it. Make sure to clearly detail which genre your project belongs in.

WHAT IF I'M CREATING A SCRIPTED SERIES OR REALITY SERIES?

If you are producing a scripted show or reality program your treatment should explain the conflict and resolution of your project.

What is the conflict? How do they attempt to resolve it?

Describe the conflict and resolution the viewer will witness. Where is the drama? What is the drama? People love drama. Remember, a basic element of life is drama. The essence of life is conflict. Life is about goals, working to achieve them and overcoming obstacles that prevent you from getting there.

It's been the key to great storytelling since *Red Riding Hood* and *Jack and the Bean Stalk*. The protagonist (good guy/girl) wants something and the antagonist (bad guy/girl) wants to prevent them from getting it.

WRITE A SHOW SYNOPIS

One page of your treatment will include the show synopsis. A show synopsis is a brief one page write up of the show idea. It explains the theme, setting, plot and focus of each episode. This is the long version of your log line. It's where you get to fully describe the show.

CREATING AN EPISODE BREAKDOWN

If you're creating an episodic show with different character's, make sure to provide sample outlines for each episode. Just a few sentences explaining the main story line of the show.

When making a show for a specific network check out some of their top shows. Research how many episodes are in a season. This is easy to do. Just go to the 'on demand' page of your favorite show and look at how many episodes there are in each season.

Once you do that you'll have an idea of how many episodes you need to create. Companies like Netflix often have shorter episodic seasons than networks like ABC or NBC. Most networks will request eight to thirteen episodes per season.

The key is to make them ask for more by bring your idea to life. Remember, you know everything about your project. You've been walking around with the idea and have molded it like pottery. No one knows as much about your project as you do. This is the first time they're getting to look at your amazing concept. Keep that in mind and treat the situation accordingly. Give them everything they need to maintain excitement about your project whether or not you're in the room.

GET FEEDBACK ON YOUR TREATMENT

Before you go running off to network executives with your show treatment, create 2 or 3 different versions. Get feedback from professionals that you trust. This will help you eliminate some of the questions people might have as you share your pitch. Many times, these questions will cause you to update the treatment. Editing is part of the process. Every time you edit you'll make crucial improvements. You want to make sure that your treatment is the best it can be.

Make sure to include the authors name on the treatment. This is the same name you will use to register your show with the WGA. We'll discuss that in further detail later in this book.

CREATING THE LOG LINE

The log line is a one or two sentence description of your project. It's like the detailed summary you see in a cable or movie guide. This logline will tell everyone what your project is about. This is what you look at when swiping through movies on Netflix, Hulu or Amazon. The short description you read when you're looking for a new movie. Some people also call this your 'pitch'.

Your logline should be ironic, that's important. As the guru Blake Snyder says, "a logline is like a good book; a good one makes you want to open it, right now, to see what's inside." When creating a logline be mindful that you should be able to get an idea of the entire project from this breakdown. The logline creates interest, helps projects get funding and provides your project with excitement.

CAN YOU PROVIDE SOME EXAMPLES FOR US?

Sure. Check out some of the loglines below. After you finish reading I would like you to create 3 loglines. One for your project, another for your favorite movie and another for your favorite book or video game.

Show: The Good Doctor

Logline: Based on a South Korean format, the drama centers on a young surgeon with Savant syndrome who is recruited into the pediatric surgical unit of a prestigious hospital. The question will arise: Can a person who doesn't have the ability to relate to people actually save their lives?

Show: The Mayor

Logline: When an outspoken, idealistic rapper runs for office as a publicity stunt and actually gets elected, he surprises everyone (including himself) when he has a natural knack for the job and slowly transforms City Hall.

Show: Instinct

Logline: A former CIA operative (*The Good Wife's* Alan Cumming) who has since built a "normal" life as a gifted professor and writer is pulled back into his old life when the NYPD needs his help to stop a serial killer on the loose. Based on the soon-to-be-published James Patterson book.

Show: Wisdom of the Crowd

Logline: It revolves around a tech innovator (Jeremy Piven) who creates a cutting-edge crowd-sourcing hub to solve his own daughter's murder as well as revolutionizing crime solving in San Francisco.

Show: Black Lightning

Logline: Jefferson Pierce (*Hart of Dixie's* Cress Williams) made his choice: he hung up the suit and his secret identity years ago, but with a daughter hell-bent on justice and a star student being recruited by a local gang, he'll be pulled back into the fight as the wanted vigilante and DC legend — Black Lightning.

Show: Life Sentence

Logline: When a young woman (*Pretty Little Liars'* Lucy Hale) diagnosed with terminal cancer finds out that she's not dying after all, she has to learn to live with the choices she made when she decided to "live like she was dying."

EXERCISE

Your task is to create 3 different loglines. One for your project, another for your favorite movie and another for your favorite book or video game.

CHAPTER 6: PROTECTING THE IDEA

This chapter is all about protecting your creative idea, your baby! First, let's have a quick conversation about creativity and ideas.

The more you lock your idea in, the better chances you have of protecting the concept. If you have a simple paragraph or one sheet about the idea with nothing else, you can protect, it but someone might have thought of a similar idea.

Upon this point I must be absolutely clear. As a writer or creator of a project you must be clear about what you're presenting. You must understand the difference between what is known as a 'generic idea' and a 'fully developed' flushed out show concept. That is why the show treatment we discussed earlier is so important.

When someone has an initial concept that is vague, generic and not fully flushed out then it is not considered protected as an "Intellectual Property". That is why my book focuses on fully developing your show idea because the more specific and original you are, the more protection you have.

If you create a show treatment, log line, lock in talent and shoot a sizzle reel or trailer, you have a solid proof of concept. That proof of concept strengthens your leverage and establishes ownership.

For example: Someone can tell you all about their amazing idea to start a rap group with a bunch of rappers who live in Staten Island. They can tell you about their plan to use martial arts samples, dark melodic beats with great samples and creative lyricism. Or, that person can play you songs like "Protect Your Neck", "C.R.E.A.M" and "M-e-t-h-o-d Man". Songs they produced, recorded and developed with a clear plan of what they wanted to accomplish. The latter will give you a much clearer picture of the concept. Often when you devote that kind of time and energy to your project great things happen. That's because you've done everything possible to make it a solid idea. You believe in the project, you've committed to the project and you are devoted to seeing it succeed. Personally, I'm happy that I was able to sneak a Wu-Tang analogy into this book. Making Staten Island proud! For anyone who doesn't know the Wu-Tang Clan, the RZA and the killer beez, please google them. Thanks. It will make sense once you do.

If you plan on making progress and building relationships in the entertainment industry, you must understand that success is built on collaborations. Unless you are the producer, director, writer, camera operator, audio technician, talent, financer and distributor all rolled into one, you'll need to collaborate with people.

That means that you'll have to share your ideas and expose your creations. Many people who are new or inexperienced in the industry are hesitant to share their creative because they are worried about someone stealing their idea. I totally understand, you don't want anyone to steal your creative 'baby'. But you cannot move forward or make any progress unless you are willing to share ideas.

The best way to stay creative is to get ideas out, bring them to life and make room for the next one. Very few creative people staked their whole existence on one idea, it's their collection of ideas that made them rich, famous and respected.

Also, people who are legitimate have a lot to lose by stealing ideas. It puts them at risk professionally, destroys their credibility and risks their livelihood with potential lawsuits. I'm not saying it doesn't happen, because it does happen. But think about this, it's much better for an executive to maintain a positive relationship with a writer/creator who can consistently bring them good ideas. That's more important than risking the relationship for one idea that may or may not become a hit in the market place. Therefore, you should only deal with companies and executives who have a proven professional track record. When dealing with outside producers, companies and networks please keep a paper trail of all correspondence. That way you have a record of all interactions.

Some legal advisors I work with believe that you shouldn't make unsolicited pitches or submissions to networks because the company receiving has no responsibility or obligation to "ideas" they did not ask for. That is because the ideas were not submitted under pre-established guidelines with the use of an 'Material Release Form'. If you have a legal team, please refer to them for advice.

If you plan on meeting with a network of production company to pitch your project, there is a document you should be aware of called an NDA or a non-disclosure agreement. The NDA is a legal contract that helps protect both parties. The agreement states that neither party can share information about the project. If you want more information about an NDA, contact an entertainment attorney.

A great plus for content creators is the ability to create a show, put it on a digital platform like YouTube, promote it and wait for the networks, agents and distributors to come to them. It's no longer a one-way street with writers and producers looking for a deal from networks. Now creators have options, different platforms where they can build audiences. Therefore, creative people have a lot more power.

HOW DO I PROTECT MY SHOW IDEA?

If you want to protect your show idea you can register it with the Writers Guild of America. As the Writer's Guild site states.

"The primary purpose of registration is to establish the completion date of your original work. Registration is available to anyone worldwide. Registration is legal, valid evidence and, if necessary, the Registry will submit material as evidence to any official guild or legal proceeding regardless of location or membership."

To register with the Writers Guild, visit WGA.org

EXERCISE

Visit the WGA.org website

CHAPTER 7: CREATING A RUN OF SHOW (ROS)

The skeleton or the frame which you build your house upon is your Run of Show. No matter the genre every production requires a Run of Show. Whenever you are producing a show or event it's something that you must have to make sure that your production run smoothly. If you want to avoid huge communication mistakes, timing issues and logistical errors, you need to create a run of show.

A Run of Show is a document that serves as a blueprint, a step by step breakdown of your entire production. For the rest of this chapter we will refer to it as the ROS – the Run of Show. This is your guideline. Whenever someone picks this up they understand exactly what is going to happen from the beginning to the end of the production.

WELL, I'M A MASTER OF IMPROVISATION. WHY DO I NEED A RUN OF SHOW?

The ROS is for the creative team. It's a document that tells all behind the scene players what to do. Each segment of the show is broken down so that producers, talent, directors and crew members know what to prepare for. Also, it's much easier to improvise creatively when you understand the structure of the show and what you're looking to accomplish. Shows like Curb Your Enthusiasm film a whole episode without a script. All they need are the scenarios listed in their Run of Show. This is a scene by scene breakdown of your entire production. It will answer some of the following questions:

What's the location?

How many people will be on camera?

Do you need any special props?

Do you need special wardrobe for this scene?

Does a special guest appear on screen with you?

What kind of props do you need in this segment?

How many cameras do you need for this segment? How many mics?

A ROS provides answers to those questions and more. It is an important document that helps everyone on the production team.

It's the road map for a successfully executed program. When producers don't create one, it's the main reason for a sloppy unprofessional production. It's the best way to master your execution.

SOUNDS COOL BUT I'VE NEVER DONE ONE BEFORE...TELL ME HOW.

If you want to create a ROS, all you need is a piece of paper. If you want to look professional, then you'll need to use writing software like MS word. People who are more experienced can use MS Excel, it's better for mathematical timing.

You need to create a table or list with the following column elements:

Ite	Scene / Element	Talent	GRX /	TRT

m #			Props	(Time)
1	Welcome to the show / Tease topics	Host	Show open Name	1:00

Item # + Show Element + Talent + Additional Elements +

Length of time

These are the basic elements, feel free to add any additional sections you'll need for your production.

When I first started in production, I was a script writer and all my shows at MTV were produced off the script. I created the script from a timing sheet that told me how many acts (segments) were in my one-hour show. It also displayed the commercial breaks. As a writer and associate producer my priority was to identify whether we were going to break or coming back from commercial. After that the creative of the segment was up to me. It was extremely fun and a great experience for a young producer/script writer.

The first time that I realized the importance of a ROS for my show was when we created the show 106 & Park: BET's Top 10 Live. As a producer there were so many elements in the show that it would be impossible to have the production team and crew to simply follow the script. The timing of the show was incredibly important because we always had to hit commercial breaks. As you know, commercial breaks are about money and no producer wants to mess up the money. The Run of Show allowed us to break down when guests would appear, how long they would be interviewed, and which videos were coming up. It also helped us plan and execute popular show segments like Freestyle Friday and Wild Out Wednesday. The Run of Show was our daily map. Getting to our destination would've been impossible without that show map.

The ROS is a vital part of the production process. The show script is usually created based on the ROS, it's a required element that every show writer needs. Every show has a ROS. If you don't have one, it's time to start making

one.

CHAPTER 8: SCRIPT OR NOT TO SCRIPT – THAT IS THE QUESTION.

Before we get into whether or not you should write a script, let's have a quick conversation about the purpose of a script. Essentially scripts are used to provide the actor, talent or host with lines of dialogue. They also offer directions for shooting, and give some basic guidelines for your setting, location and crew. If you are creating a movie or TV show, there's an established format for script writing. If you're writing for anything else like a hosted show, an event, a podcast or a vlog, you have more freedom and it's best to use a format that fits your specific needs. The amount of information that's included in your script is up to you. Typically, it will cover things like your show introduction, topics, teases, dialogue, guest questions, cues and your show closing.

If you have followed the steps laid out in the previous chapters, you already have a basic story and plan in mind. You have a log line, a synopsis, an episode breakdown, characters, settings and some episodes roughly planned out.

In addition, you should have a format with all of your scenes listed. Now it's time to create the script!

The script should follow the format. Just write. Don't worry about mistakes. The key to writing is to get it all out. A writer always must go back and edit their work. Writing is all about editing. Reading what you wrote and making it better with each revision. Sometimes there are multiple changes, sometimes there are just a few changes. Enjoy the process. Nothing feels better than a completed script or book.

When working on your first drafts, keep in mind, you can always go back and edit, especially when you aren't sure of which words to use. You can add more action or dialogue later, just keep writing. When you aren't sure about dialogue, bounce ideas off of friends or another creative peer. As discussed before, production is truly about collaboration. You'll be amazed at how people can inspire you when you're developing and creating your project.

One of my favorite things to do when writing is to read the dialogue myself, in the same style and tone as the performer. If someone else is around, I might ask them to play a part while I play another part. There's a major difference between the two. Words and dialogue sound different when read in your head versus when it's being read aloud. Many times, that's when I really get to fine tune my script. It's a fun process and I enjoy doing it. This exercise allows the writer to fully understand why some edits need to be made. It brings an element of realism and fun to the process. You'll be able to tell if there are too many words or if things sound 'unnatural'. Fix those issues immediately. It's best to do this before your talent reads the script.

I HEARD ABOUT TABLE READS FOR FILMS. SHOULD I DO A TABLE READ?

I'm a fan of talent table reads because they help you see which parts of the script your talent is comfortable or uncomfortable with. For those who don't know, a table read is when your show talent reads the script at a table before getting on set to perform. This is a small team session, normally with talent, show producers and writers who are there to make any adjustments and answer any questions talent has about the show. If they have any questions or concerns, the production team can address them immediately. Often talent will make slight changes, you always want your talent to be comfortable. They are the ones who will appear on camera. You need to make sure it works. But, if you feel strongly about a word or phrase, make your point but don't argue.

When writing scripts for people like Dave Chappelle, Bill Bellamy, Jonah Hill, Channing Tatum, Kevin Hart, Matt Pinfield, Tyrese and other celebrities I learned the art of letting talent ad-lib. This is especially important when working with comedians. They will take your words, include your direction and deliver it in a way that's fully natural for them. Often you get the best product this way.

HOW CAN I LEARN MORE ABOUT SCRIPTWRITING?

If you're writing a film script, you should read screenwriting books like Syd Fields classic book: Screenplay, The Foundations of Screen Writing. The book includes industry standard rules such as where and how to place names, action, settings and other important elements.

WHICH SOFTWARE SHOULD I USE?

The industry standard software for script writing is Final Draft. It's a great program and has numerous tools that make the script writing process easier. Programs like Final Draft were created to make movie scripts. If you are writing a script for your show, podcast or vlog, a program like Microsoft Word is all you need. You can use any program that lets you type out your ideas and share them with your team.

Now you have the tools to start writing your script. I can't tell you what to write and where to write it because it's your script, your story. So, have fun and start writing. I look forward to seeing your project come to life. You wouldn't be reading this book if you weren't devoted to doing it. Also, if you have a budget and don't have time to write, you can always hire a script writer.

EXERCISE

Take one scene from you project and write the script. Just one scene.

Then read it and make edits the next day. Have fun.

CHAPTER 9 LOCKING IN TALENT

If you are a producer who plans on selling your show to a network, it's important to have talent attached to the project. There are some projects that get green-lit without it but it is rare. Having talent attached helps improve your leverage. It increases the possibility that the network will keep you attached to the show and let you produce it. Especially if you have no prior experience producing a show for a network.

If you're planning to pitch a reality show about a specific topic, make sure that you have talent that fits the genre. Unless it's a show about your life. If it's about your life, then it's safe to say, you already have the talent confirmed and available.

WELL, I'M TALENTED! WHY WOULDN'T THEY GIVE ME A SHOT?

Many networks have relationships with producers and production companies they like to work with. Usually they are proven professionals who know how to deal with network demands, handle production teams, deliver shows on time and on budget.

When you go to the network you are asking someone to invest in your show idea. All risk is on the network. If you only have a concept without talent attached and no prior track record producing network shows, it will be easier for them to say no. That doesn't mean it can't happen, anything is possible. It just means that your chances will be slimmer depending on the network.

DO YOU MEAN I NEED A FAMOUS ACTOR, ATHLETE OR CELEBRITY?

This doesn't mean that you'll need a famous actor, athlete or celebrity. You just need to have someone compelling and interesting enough for your show. That person should match the concept and bring value to your project.

In the world of reality tv you don't have to be a celebrity. But it helps if the person has a strong story or a strong name in their industry. Think about some of the shows on networks already. Most reality shows don't require an A-list Hollywood celebrity, just great personalities who make the show entertaining to watch. As a producer you're spotting talent, bringing that talent to a wider audience and creating a star through your show. Go get 'em!

I DON'T CARE ABOUT NETWORKS! I WANT TO MAKE A SHOW FOR YOUTUBE.

Here's the positive. If you are producing your show for a channel like YouTube and you're the talent, then you have a huge advantage. You are developing your fan base. You are building your audience. When the audience is big enough, you can leverage your fame (that you've worked hard for) into other opportunities. A perfect example of this is YouTube superstar Lilly Singh aka Superwoman. She has an inspiring story, please read her book. A quick summary is that while down and depressed she started her own YouTube channel. It started off with funny videos starring family and friends. In time she built a strong fan base, mastered her craft and now she's a well-paid celebrity who has done videos with Will Smith,

Michelle Obama, Seth Rogen, Dwayne "The Rock" Johnson and many others. She also released a best-selling book and went on a world tour. Yes, that is possible! So be encouraged, start small and build your own fan base. The rewards could be huge. It's always good to bet on yourself!

WHAT IF I JUST WANT TO SELL MY IDEA TO A NETWORK?

If your plan is to come up with a great idea and sell your idea to a television network, there are a few things you should keep in mind. It's difficult for new producers to sell a show format or genre they've never executed before. It's especially difficult to pitch a genre without talent attached because normally there's a production company or producer who is more experienced and successful at producing the genre.

That's why you see so many independent filmmakers writing, producing and directing their first films. After the success of that initial film or television show, sky's the limit!

Just ask the guy who wrote, directed and produced *Paranormal Activity*.

Normally someone who develops a hit show becomes popular in the industry. Once they have proven that their style works, they become the go-to person in the industry for that kind of content. Their name has cache because it's attached to a hit. Everybody in the industry loves a hit, especially one that made money.

Those individuals usually become leaders of the genre. Once they establish themselves as a leader in the genre, they dominate it until someone else creates a successful project that can compete with them.

THAT SOUNDS GOOD, BUT DO YOU HAVE ANY EXAMPLES?

Here are a few examples:

Mark Burnett: After emigrating from the UK, he worked as nanny and sold t-shirts on the beach. After some time, he decided to launch a career in TV. The first show he developed was called *Eco-Challenge*. That show launched his career as a producer. After creating *Eco-Challenge*, he was inspired by the Australian television series *Expedition Robinson* (an example of a new twist on a previous idea) and created the hit reality show Survivor. After Survivor became a hit he went on to produce *The Apprentice, Shark Tank, Are You Smarter than a 5th Grader?* and *The Voice*.

Shonda Rhimes: Shonda has an incredible story. Her career is truly inspirational because she's an example of how hard work and perseverance can take you to the top. After graduating from USC, she found her way in the industry and received an opportunity to co-write the successful HBO movie *Introducing Dorothy Dandridge* starring Halle Berry. It went on to receive numerous awards. In the years following, she co-wrote the sequel to Princess Diaries called Princess Diaries 2 and wrote a film for Britney Spears. After watching the success of romantic dramas and medical dramas like St. Elsewhere, she decided to create her own pilot called Grey's Anatomy. Her twist on the genre was a medical version of Sex

In The City that would feature a multicultural cast. Grey's Anatomy became a huge hit for ABC and launched her career as a showrunner. She later produced spin- offs series like *Private Practice* and most recently owned ABC Thursday nights line-up with the hit shows *Scandal* and *How to Get Away with Murder*. If you want a hit drama, you better call Shonda!

Michael Wolf: Started off his career as an advertising copywriter with dreams of writing films and TV shows. He became a staff writer for Hill Street Blues and Miami Vice and later created the hit show Law & Order. The show became a huge success and launched multiple spin-offs – expanding the franchise. He also created other hits shows such as *Chicago Fire* and *New York Undercover*. If you're looking to do a procedural crime series, he's the go-to guy. In this case the writer was the star, his ability to write successful shows and films allowed him an opportunity to pitch and create Law & Order.

Mona Scott-Young: This reality show powerhouse started off as a successful music industry executive. After leaving the industry, she launched her television career by producing a special called The Road to Stardom with Missy Elliott. Shortly after, an executive at VH-1 asked her to help create a show for a rap artist named Jim Jones. Instead of focusing strictly on Jim Jones she chose to focus on his girlfriend Christy and her circle of friends. The show was titled Love & Hip Hop and it became a huge success. It spawned a franchise with multiple shows and spin-offs. She executed two major strategies that we discussed earlier. She put a new twist on the popular show Basketball Wives. You could say it was a form of Hip-Hop wives. She also came to the table with talent locked in, Jim Jones. Boom, boom – pow!

If you are creating a show with a specific network in mind, make sure that the show matches the network and their audience. Do some research on the network and study the shows they currently have in production.

If you're getting ready for a pitch meeting, watch Shark Tank. It might help you. It's the closest example I've ever seen to an actual pitch meeting. Except, you have much less time than they have. So be ready, be prepared and sell that show!

CHAPTER 10 BUDGETING YOUR PREPRODUCTION

Your show is finally coming together and you're almost ready to shoot.

WHAT ELSE DO I NEED? I HAVE A TREATMENT, LOGLINE, TALENT, AND A SHOW FORMAT?

Well, as you know making a show, podcast or web series requires more than a script, talent, crew and equipment. The only way to make sure that you have enough money to pull off your great idea is to create a budget. Don't hire the crew or a production team without your budget in place.

WHY DO I NEED A BUDGET?

The budget is a guideline for expected costs. It outlines the expenses you will incur during every stage of production. Your budget will be able to tell you where you're saving and where you're over spending. In some cases, it will tell you if you are going to need more financing. A well-planned budget will help you predict and compensate for anything that might occur.

It also allows transparency if anyone is looking to fund your production. You can show them how you plan to spend the money and where their money is going. This is something you would only do upon request. Also, if you're creating a project for a business, church or school they will have a budget, a certain limit on how much they can spend. Once you know the number, you and your team can properly budget the production.

The budget will clearly show the producer and the production team where the money is going. If the team would like to invest a larger portion of the funds in a different area, they will determine where the sacrifices need to be made to accommodate that change. This is especially important if you're using 3rd party music, videos or images; these elements are often expensive and have a major impact on the budget.

OK, SO WHAT DO I NEED TO INCLUDE IN MY BUDGET?

Your budget will cover every element through pre-production, post-production and final delivery. The budget is always based on one of the following:

1. The amount of money you have.
2. The amount of money you can raise.
3. The amount of money someone has given you to complete the project.

If you are an independent creator and plan to get funding from a fund-raising site like kick starter, the budget will help you determine your fundraising goal.

PRE-PRODUCTION ELEMENTS

Film / Video Crew

This is the time to think about what kind of crew you're going to use. If you're filming a pod cast or single camera video series, you might be able to find someone who will help you and work for whatever rate you have. If it's a larger production, you need to decide whether your crew will be union or non-union.

TALENT

If you have talent attached, it's important to determine what kind of fee they will require. If you have actors, you will need a contract and you'll probably need workmen's comp insurance.

EQUIPMENT

You need to determine what kind of equipment you plan on using; this includes lights, camera equipment, crew size, sound equipment, gaffer equipment and grip equipment. Don't forget about other necessary elements such as hard drives, tape stock, batteries and other ancillary needs. Make sure to discuss equipment with your crew and production team.

PRODUCTION & FILMING

Before you start filming make sure that you have everything you'll need to film the project. That means that you should be prepared to account for location permits, security fees, van rentals, car rentals, transportation, tolls, catering, craft services (food), wardrobe stylist, hair, make-up, art design, props, electrical fees, parking fees and any additional production needs.

POST PRODUCTION

You must plan your budget for the post-production process. This can include but isn't limited to; an editor, editing equipment, graphics, effects, color grading and audio mixing. You also need to account for any 3rd party music and images. Depending on who you use and who you hire these items can become quite expensive. It all depends on your budget and who you can afford.

OK, SO...HOW DO I CREATE A BUDGET?

They are a few budgeting tools available to creators. If you plan on having an intricate production that requires a lot of moving parts, you probably want to use a quality budget software program. Most of the Line Producers and production managers I know use a program called Movie Magic by

Entertainment Partners. It's very useful and comes highly recommended.

For many years, I simply used MS Excel to create a budget for my projects. That software is available to anyone who has MS Office. It was a very helpful tool whenever my budgets were ten thousand or less.

If you really need help creating a large budget, consider hiring a production manager or line producer. Someone with experience in this field can really help you create a proper budget.

FINAL THOUGHTS

Once you have your idea and treatment locked in you should start working on the budget. Outlining a budget will help you with some of your decisions. It's always good to discuss the budget with your crew because they might have alternative ideas about how something can be accomplished. Thinking ahead of time will let you create clear plans about equipment rentals, shoot schedules, locations, length of shoot days and all your other production needs.

CHAPTER 11: CREW UP! HIRING YOUR CREW

You're almost ready to shoot. You have a show treatment, a log line and a production budget. Now that you know what you can afford, it's time to line up your crew. Your team is normally determined by the size of your budget. When assembling your crew, it's important to make sure that they realize your vision and what you're looking to accomplish. It's especially important if you have a limited budget. That normally means you can't afford to pay people what they are normally paid. They are doing you a favor. Keep that in mind and respect their choice to join you.

A production crew is all about creative collaboration. You want a leader in every department. Each member of the crew should come to the production with their best abilities. That often means listening to different perspectives and being open to suggestions. It's your show but a good creative producer makes the experience rewarding for everyone involved. Your team is there to support your vision so make them feel good about it whenever you can. At the end of the day, everyone wins when they can put a great project on their reel. Plus, nothing is better than experience. Get as much as you can.

HOW DO I GET STARTED? I DON'T KNOW ANY PRODUCTION PEOPLE

First, analyze your show, treatment and budget. Determine the amount of people you'll need to hire for your project. Break down your project scene by scene. Make sure to detail location needs, wardrobe needs, equipment needs and any other production needs.

If you don't have any shooters, camera men, audio mixers or production staff in your immediate circle of friends, there are other options. You can let people know that your production is hiring by using sites like Mandy.com, Craigslist and ShootingPeople.org. You can try your nearest film school or film program; students are usually hungry for experience and might be willing to work on a project to help build up their credits.

If you know some production professionals, get advice from them. See if they can recommend anyone. They might have names and contacts available. Make sure to collect resumes, references and reels so that you can properly research your team.

Vimeo is a great resource for production reels and camera tests. Professionals upload footage using the latest equipment, take some time out and view it whenever you can. It will help you when thinking about options for your project.

It's easy to get stressed out on a production set. Do your best to keep a positive atmosphere. You might be the creator, the talent or the director but without the crew you don't have a production team. Keep them inspired and treat them with respect.

EXERCISE

Start breaking down your project. What will you need for your shoot?

How many cameras do you need?

What are your audio options? Do you have a boom mic or Zoom recorder?

Do you need lights?

Do you have a location?

CHAPTER 12: SHOWRUNNER & LINE PRODUCER

Let's have a quick discussion about Showrunners and Line Producers.

SHOWRUNNER

A showrunner in television terms is normally the lead executive producer of the show. That person has creative control of the production. They juggle the responsibilities of being the head writer, executive producer and story editor. They develop storylines, hire crew members, deal with network executives, write scripts, cast actors and manage budgets. A showrunner is responsible for all creative aspects of the show.

The person in this position has been selected to run the show because they've been approved by both the production company and the network. I know a little about this position because I've been hired as a showrunner by different networks. In the film world the top creative position is held by the director but in television the top spot is held by the

Showrunner. If you hire a showrunner, this person is responsible for delivering a finished product.

WELL, HOLD ON…ISN'T THE CREATOR OF THE SHOW, THE SHOWRUNNER?

Good question! Sometimes that is true but sometimes it is not. Typically, the showrunner is the creator or co-creator of the series. If the person who created the show has no experience producing television shows or minimum experience managing a creative team, the network will hire an experienced showrunner to handle those duties.

LINE PRODUCER

The Line Producer is the head of the production management team and the person who manages the budget line by line. The person in this role is over the money, payroll, daily schedules, crew schedules, logistics and locations.

Once pre-production starts their responsibilities include assembling the production team, recruiting key crew members, dealing with crews and unions, renting equipment, setting up the production offices, lodging, organizing the shoot schedule, monitoring the budget, handling budget needs, catering, casting, managing release forms along with safety and risk management.

The Line Producer is responsible for keeping the show on budget. An effective production always does it's best to stay on budget. If you stay on budget and deliver a great product, you will keep working. Unless you're Steven Spielberg or James Cameron. But if you're going in that direction make sure your blockbuster projects steadily earn between $100 million and $2 billion each.

Many people relate this position to the chief operation officer of a company. The Line Producer works directly with the showrunner. These are the top two positions on the production team. They need each other and will work in tandem until the end of production. Every team decision made from pre-production to final delivery will go through these two department heads.

CHAPTER 13 RESOLUTION OPTIONS. PREPARE FOR YOUR SHOOT

Now it's time to talk about video resolution. In this chapter we will discuss some of the top video resolutions available so that you can decide which resolution would be best for your project.

Overall, the term video resolution means the quality of your video image. The resolution is the amount of detail a video image holds, in technical terms, the number of pixels per unit of area used to create a video image. Higher resolution means more pixels and more detail in your image.

If you are creating a project for a specific network or video streaming service, make sure to find out if there's a video resolution that they require. Some companies require a minimum resolution of 720p. But as you know, when shopping for televisions, computers, smartphones and digital devices; there are better resolution options available like 1080 and 4k.

If you want to keep your project up to date and sellable make sure to use an HD resolution. If you don't, competing with other shows and video content will be difficult. Today's consumer is used to high quality digital images. Always think about your consumer or viewer. What would you want to watch? What kind of video quality do you, your friends or your family watch? If they have an iPhone or a smart phone, they want a high quality digital image.

Now that we've had a brief conversation about why you would want to use higher video resolution options, it's time to discuss some of the best options currently available.

720p - This is short hand for 1280 x 720, known as basic HD or HD ready. This is the lowest level HD video resolution you should use for video content.

1080p - This is short hand for 1920 x 1080, known as (FHD) or Full HD. Currently, this is the resolution most networks utilize.

1440p - This is short hand for 2560 x 1440, more commonly known as QHD or Quad HD. You might see this resolution on gaming monitors and high end smart phones.

2160p - This is short hand for 3840 x 2160 better known as 4k, UHD or Ultra High Definition resolution. 4k gives you an amazing video image. It's a large display resolution found on high end TV's and monitors. 2160p is called 4k because it is four times the resolution of 1080p (FHD) which means it packs a serious punch, of resolution. When filming in 4k keep in mind that it requires a large amount of storage space, especially when filming on data cards or storing footage on drives. Also, make sure that the network or provider you're working with supports 4k video resolution.

4320p - This is short hand for 7680 x 4320 also known as 8k, it offers 16 times more pixels than the regular 1080p (FHD) resolution.

Review this chapter whenever you're selecting resolution for your project. Always keep in mind, higher resolutions require more hard drive space to store footage. 4k footage will require more storage space than 720 and 1080 footage. Just something to think about when making your decision.

CHAPTER 14 FRAME RATE OPTIONS

Let's take a moment to discuss frame rates for your show.

When filming a show for a network or website you should contact them and ask them about their show deliverables. Most networks have a document that breaks down their digital delivery requirements. Sometimes it's called a 'tech spec sheet' which is short for 'technical specifications' sheet. Once you receive this information, share it with your team, let them know what the requirements are. It is vital that you communicate this information to your production team including camera operators, director of photography, audio, editors and post staff.

Choosing the proper frame rate is important. It will impact the look and cinematic feel of your show. You must think about the right frame rate, resolution and camera for your project. The frame rate and resolution you choose will also affect your delivery process. Some of you will have creative license, you'll be able to choose what works best for you. Others delivering to a network or streaming service like YouTube must research what they require. Make sure that you deliver to those specifications.

If you are not able to obtain this information prior to filming all is not lost, there are a few options for you. This chapter will explain some of the popular video production frame rates. It's about to get technical!

Let's review some of the most common frame rates in production:

Note: when you see *fps* that means *frames per second*.

30 fps / 29.97

Some networks can broadcast video recorded in a 29.97 or 30 fps frame rate. When people say 30 fps the actual technical breakdown is 29.97 frames. In production that number is often rounded out to 30 frames per second. It's easier to say and write. This is a safe frame rate to use for most networks. If you're dealing with a specific network, please check with them to make sure that this is their preferred frame rate.

59.94 / 60 fps

In recent years I have worked with networks who have asked us to deliver shows at 60 fps or 59.97 fps. This is a safe frame rate for most networks and provides a very clear image. The image has no stylization, many professionals say that it looks 'like video'. This frame rate is also popular in video games. One of the things it does is minimalize motion blur, keeping the image clear when there is fast motion on the screen.

Once again, if you are delivering to a specific network make sure that this is a frame rate they'll work with.

24fps / 23.97

This frame rate is for people who want a specialized look. Many call this the 'film look', because it's the frame rate used when filming movies. Once again, the frame rate is called 24 fps, but the actual breakdown is 23.97 fps. When prosumer digital cameras were first released one of the coolest things you could have was a camera that gave you a 24p look. Due to advancements in digital technology this is currently an option on most digital cameras. This feature along with digital depth of field has really taken the digital camera to the next level.

Now you are armed with some knowledge about frame rates for your project. Make sure to use the best one for your production needs. Always discuss frame rates with your camera team, editor and post production team. If you are creating video content for a network, I must reiterate, make sure that you find out which frame rate works best for them. If you really need someone who dives deeply into this, hire a technical advisor for your project.

CHAPTER 15: SIZZLE REEL, PILOT, TALENT REEL OR SERIES?

It is important to decide whether you are going to make a sizzle reel, talent reel, pilot or multiple episodes. The answer relies on different factors. How much money do you have to shoot your project? What are you looking to do with the final product?

It's your decision. But keep in mind, you can make a sizzle reel out of a pilot or series, but you can't make a pilot or series out of a sizzle reel. Make sure that you think about your strategy before you shoot. This way you have a strong game plan for the footage in post-production. While working as an executive producer at Viacom, I've been there to see people get their shows green-lit multiple ways. Some people had a great concept with talent attached. Others had a show treatment and a sizzle reel. Other people brought in pilots that showed promise, those pilots were developed into a series. You must prepare yourself with a strong concept, a great strategy and talent that fits the network. In my experience, it's often easier to get a network executive to watch a sizzle reel, video clip or trailer, especially when you have no prior relationship with them.

<u>CAN YOU BREAK DOWN THE DIFFERENCE BETWEEN EACH ONE?</u>

No problem. Here's a quick breakdown of each option:

TALENT REEL:

A Talent Reel is a video that demonstrates why your talent would be a perfect fit for the show. It's normally just a few scenes or a pitch video with talent talking straight to the camera. Talent reels are often used for reality shows because the talent is critical to the ideas success. You want to make sure that it showcases their personality and proof of why the concept would work with them in the role. You also want to provide an example of what will make them interesting to watch and why they're a great fit for the show. In my experience popular celebrities often use talent reels to pitch themselves for projects. Many times, it shows networks that they are committed to the producer, the production and fully involved with the project. The actor Ken Jeong from The Hangover fame used a talent reel to pitch his show Dr. Ken to ABC. It worked and they green lit the series. Ken Jeong used a mix of his star power, commitment to the project and proof of concept to get the job done (he graduated from Duke and earned his medical degree from Chapel Hill).

Normally a talent reel is approximately 2 – 5 minutes long.

SIZZLE REEL:

The sizzle reel is the first major visual introduction of the project. It's when you've gone from words and treatments to the actual look and style of the show. It provides a clear vision. It often looks like a trailer or commercial for your show. It even includes partial scenes that highlight strong moments. The style of your sizzle is completely up to you. It must represent your project. You should go online and look at different sizzle reels to get an idea of styles you like. There are times you'll be in the room with executives to show your reel but many times the reel is sent and they're watching it on their own. You and your team aren't in the room. So, make sure to keep it entertaining and engaging.

The sizzle establishes characters, themes and pacing. Editing is very important when creating a sizzle reel. Music is also critical to reel. I often recommend that people use different pieces of music to keep the viewer engaged and to maintain energy throughout. A sizzle must hold an executive's attention while simultaneously selling them on your idea.

Normally a sizzle reel is 2 to 5 minutes long. We usually like them to be approx. 2:30. If you have a 5 minute reel it better be incredibly entertaining.

PILOT

A pilot is a significant step up from the sizzle reel. It's a full sample episode; complete with scenes, music, graphics and spaces for commercial breaks. If you create the pilot on your own, you have full creative control. You have an opportunity to create the exact show you want to make. Pilots aren't cheap, so in poker terms, all your chips are on the table. Now I don't play poker, but I thought that sounded appropriate. If it works, you have bargaining power and lots of leverage when dealing with a network. Making a television show is a huge accomplishment, especially when it's professionally produced and packaged. You're official on another level.

In my career, working at networks and with development departments, I've been hired as a producer or showrunner to make pilots. Usually that means creating a show which was either my concept or an executives concept. Either option is fun and challenging because we get to create the show from scratch. There is no road or formula laid down for you to copy. It's a great feeling for the entire team when the pilot gets approved to air.

Pilots can be expensive, especially if you're spending your own money. Plus, there's no guarantee that the network is going to buy it. It's a creative risk and an investment that might not work out. But it's your show, your vision and your dream. There's a great saying "You miss 100% of the shots you don't take. So, you might as well take the shot." If you can afford it and believe in the project, go for it!

WHAT IF I'M NOT PITCHING TO A NETWORK?

FULL SERIES:

If you plan on creating a full series, then you're probably going the independent route. That means you are looking for a digital platform or a distribution partner. YouTube, Vimeo and Soundcloud are great platforms for independent creatives. Anyone creating a podcast, YouTube series or Vlog series can pursue this route. This allows you to hone your craft and build an audience without dealing with a network. This route has been successful for many creatives on YouTube. That is what I call betting on yourself. If you're successful, you have all the leverage. When the networks come calling you're able to set the price and control the conversation.

CHAPTER 16 SHOOTING YOUR SHOW

You've come a long way and it is finally time to shoot. It's time to execute your vision and bring your idea to life. Everyone reading this has a different show idea in mind, so this chapter was created to provide overall tips that will improve the quality of your shoot. No matter your experience level, if you follow these tips you'll be much happier in post-production editing your project.

YOU MENTIONED EXPERIENCE. WHO HAVE YOU FILMED SHOWS WITH?

These tips were gathered through my experiences creating television shows, documentaries, digital series and branded content. Along the way I've had the honor of working with top level talent like Will Smith, Denzel Washington, Channing Tatum, Beyoncé, Vin Diesel, Tracee Ellis Ross, Tom Cruise, Bruce Willis, Queen Latifah, Kevin Hart, Dave Chapelle, Jonah Hill, Jamie Foxx, Kanye West, Ava DuVernay, Issa Rae, Seth Rogen, SZA, Tyrese, Issa Rae,

Eminem, 50 Cent, Jay-Z, Common, Snoop Dogg, LL Cool J, Sean Combs, Erykah Badu, Pharrell, Nas, Ludacris, New Edition and Jill Scott. Those are just a few of the names, apologies to any celebrity I've worked with who reads this and feels like I left them out. The list is really long but I'm honored to have work with all of them. Each shoot is a new experience.

This chapter will detail a list of things to look out for on any shoot. It's important to read and utilize this list before your actual shoot day. Use it to prepare prior to your shoot and make sure to go over it with your core team on your shoot day. Following these tips will help you have a productive and professional shoot. You'll be excited about the final result.

PRODUCTION SCHEDULE

A successful shoot day is dependent on a well thought out production schedule. If your shoot is not carefully scheduled, your budget and your production will run into serious problems. Every schedule is different but they all start out with a solid breakdown of the project. The breakdown allows you to detail the shoot from beginning to end. That means call times (when people are expected to be on set), hair, make-up, wardrobe, lighting, camera set-up, scenes, locations, meals, breaks and wrapping the shoot on time so that your production stays on schedule. Your shoot breakdown allows you to schedule and properly manage every element of your production. You want to be smart, strategic and cost effective. Learning how to schedule a shoot means that you are properly managing every asset you are responsible for. All productions, whether big or small, should have a production schedule.

SHOT LIST

Now that you have a show treatment, a show format and a script it's time to develop your shot list. A shot list is a list of the visual elements you want included in your shoot. It's a check list for the producer, director and camera team. If you're an independent creator handling all those roles, this list is for you! Don't worry, I've been there too. The shot list allows you to be more organized and provides you with a breakdown of the shots you'll need to get.

If you have a larger team it ensures that the key production people are on the same page and aware of all the shots needed that day. The shot list will make your day more efficient. When creating your shot list make sure that your shoots are based on schedule, location and setup.

HOW ABOUT SOME FILMING TIPS? I WANT TO LOOK LIKE A PRO.

Whether filming with a crew or by yourself, here are some tips that will help you during your shoot.

FILMING TIPS

1. Check your white balance whenever you shoot.
2. If you have multiple cameras, make sure they both white balance.
3. All cameras must check and match their settings; frame rate and resolution.

4. When you are filming outdoors make sure that the sun is facing talent. Not behind the talent. Let the sun light your subject or scene.

5. Pay attention to the composition of your shot. Don't cut off anyone's head by mistake or leave too much headroom above talent.

6. Hand-held shots are great, but only when they are stable. Unless you're going for the Blair Witch shaking camera shot, only move the camera when necessary.

7. If you are the camera operator and the main talent, use a tri-pod.

8. If you are recording a lengthy event, concert or production, use a tripod.

9. If you are recording someone talking straight to camera, use a tripod.

10. Make sure to get b-roll and cut-away shots for your edit. It helps.

HOW ABOUT FRAMING, COMPOSITION AND BACKGROUNDS? DON'T FORGET ABOUT THAT.

Make sure that you always study the composition and background of your shot. Early on in life, my love for comic books and visual art helped me understand how to compose images on a blank page. I loved to draw. I loved to paint. Your camera frame is your blank canvas. You are the person filling that frame with whatever you want your audience to see. It's all under your control. Remember, the audience can only see what you put in the frame. The content, the angle and the action, is all up to you.

Think about what's in your shot and how your shot is framed. Try to make the shot interesting and compelling. Make sure that you focus on your subject. Never let your background be a distraction unless you intend it to be. This is your frame.

As a child when I read comic books I always noticed how my favorite artists at Marvel and Image comics created incredible pages with their masterful composition. Artists like Jack Kirby, Frank Miller, Jim Lee and Todd McFarlane do a marvelous job (Marvel pun intended) creating action packed compositions. Chris Nolan did the same in the Dark Knight film series, Patty Jenkins in Wonder Woman and Ryan Coogler in Black Panther. Of course, people like Martin Scorsese, Steven Spielberg, Oliver Stone, Kurosawa, Hitchcock, Godard, Spike Lee and James Cameron are masters at it as well. It's all about focusing on the action, focusing on your talent and making sure your scene moves your story forward.

WHAT ABOUT ITEMS IN THE BACKGROUND? DO I HAVE TO WORRY ABOUT THAT?

That's a fantastic question. If you are looking to sell the show, always avoid unnecessary brands, labels, artwork and imagery that you do not own. Whenever you are dealing with networks or distribution companies they are required to clear, license and pay for any background imagery they do not own. If you are an independent creator, the last thing you want are huge blurred out images in your show. Especially if it's because someone couldn't pay for clearance. One example is when I was a producer for Season 6 of ABC's popular show Wife Swap. We had a family shopping in a grocery store and the legal department wanted us stay away from certain brands. So, we had the art department make fake boxes of brands and we turned other boxes around while filming. We skillfully placed tape over labels so the branding was covered. We did the same thing when filming in a family's kitchen. This is something I've had to repeat on other documentaries I produced and directed. Make sure to scout the location and have some black tape ready. You never know if you'll need it. Just remember; If you don't own it, don't shoot it unless you can pay for it.

THINK ABOUT THE EDIT

Always think about how the video will be edited together. The best shooters, producers and directors are always thinking about how their shots will work in post-production. Make sure to get enough pre-roll and post-roll for the edit. Don't immediately stop filming when the person is talking. Wait another 5-10 seconds. Make your editors life easier. You want to be the producer who doesn't come in with poorly shot footage, bad audio and not enough b-roll.

DO YOU HAVE ANY TIPS FOR A PRODUCTION USING A SCRIPT?

If your production is using a script, make sure that you have enough copies on the set for everyone involved. Before you make copies, it's important to make sure that you have the most recent version clearly marked. The most recent version is the one that should be on your set. So, if you are on your third version of the script, it should say version #3. When producing award shows or live shows we often have the script team make the shoot day script a specific color. That way it is easily identifiable and different from previous versions.

LIGHTING

Good lighting is important when dealing with image quality. There are some cameras that work well in low lighting but the last thing you want is a grainy shot. When filming, try to get a light kit with at least three to five lights in it. It will make a huge difference in the quality of your shots.

A basic three-light set up would include the following: A key light, placed close to the camera to light your talent. A fill light, aimed at the subject and set up on the other side of the camera. And, a back light placed behind the subject to separate them from the background. Make sure that the stand for your back light is out of the camera frame. If you have more lights use them to light the background and use colored gels to brighten things up. Lighting the background creates separation between your talent and the background. It also adds depth to your shot. You always need distance between your subject and your background.

If you are filming outdoors on a sunny day, make sure that the sun is facing talent, not behind your talent. Use a reflector if you need some extra light.

GREAT TIPS FOR VIDEO, BUT I HATE LISTENING TO BAD AUDIO WHEN WATCHING VIDEOS. ANY WAYS TO HELP?

Here's a list of five (5) things you should be mindful of when recording audio.

1. Use an external microphone if you have one. Don't rely on the camera microphone for great audio.

2. Keep your microphone close to talent.

3. Have a producer or team member listen to the audio through headphones while you shoot and record audio.

4. Use a hand-held mike for (MOS) man-on-the-street interviews. Wireless mics are usually the best option. Try to keep cables to a minimum.

5. Use a lavalier or boom mike for interviews in a studio or inside location.

FINAL THOUGHTS

In production, we get paid to make sure things go right but we are re-hired and appreciated because we know how to execute when things go wrong. We always want the best, but you must plan for the worst.

A major part of pre-production is planning wisely and preparing for anything that can go wrong. A tight schedule is understandable, but you never want to cram too much into one day. Make sure to plan and schedule your day properly. Never forget to give your crew breaks, meals plus water and snacks. Keep the crew happy and stay positive. A production shoot is an experience everyone on the team will remember. You are bonded for life.

CHAPTER 17 POST PRODUCTION

You are a champion and you should feel good. Why? Because you're at the third and final stage of the production process. Just to re-cap, in case someone skipped ahead, the three stages are: pre-production, production and post-production. You have completed a major part of your goal and you finally have footage of your show. You molded your idea, created a treatment, a logline, a format, a script, a budget, hired your team, locked down locations and led your production. Kudos to you!

Next major step is to make a show out of the footage you shot. The main goal is to keep the best moments and eliminate anything that doesn't work. Once you decide on the footage that works best, put together your show and enhance it with sound design, b-roll, pictures and graphics. Some might find the process tedious but it's fun. Each layer enhances the overall project, enjoy the process. Some of you will choose to hire an editor. Others will choose to be their own editors. Either way, these steps will help you during the post-process.

WHICH SOFTWARE SHOULD I USE TO EDIT?

EDITING SOFTWARE

It's always good to select your editing system before shooting principal photography. If you didn't, no problem. There are three primary software options, you must decide which one is best for your needs. The options are Adobe Premiere, Avid and Final Cut. Find out which software would be best for your project. If you already own one of them, and that's all you can afford, then decision made. Another factor is deciding which system your editor works best on. Some editors have preferences and you should always try to match your editor with their preferred edit system.

EDITOR

Hire a good editor. Or, hire the best editor you can afford. The editor will read your script, look over your format and screen all your footage. After doing this they will begin putting the show or story together based on that information. Some producers like to be present during that process while others like to check in.

I always like to decide on an editor before the project goes into production. This way they can make suggestions or add any creative input before you film. A good editor will advise on the types of shots you will need for post, along with any other post-production thoughts before the filming starts.

The length of time your editor is on the project is dependent on your budget, rate and your agreement. In the process of editing, your editor will create different versions of the show, making critical updates on every version. Usually the first version is an Assembly edit, then a Rough cut, a Fine Cut and a Final Cut. The Final Cut is the term for the final edited version of the show. Now here's a tip for you to look like a pro. There are two conclusions to an edit. The first is when all visual images (aka video) are approved and finalized. The second lock is the audio lock or sound lock when the show is fully mixed with all sound effects and elements included. If you are sending your audio out to a 3rd party vendor to get mixed, you cannot do a proper sound mix until video is locked.

AUDIO MIX

Adding music truly depends on the type of show you're producing. If you plan on adding music to your project this section is for you. Once you have a tight edit of your show you need to enhance the video with sound and audio. For this process you will need an audio mixer. The audio mixer will work with an audio mixing or 'sweetening' program that will make your project come alive. They will enhance dialogue tracks, add or enhance sound effects and mix down the entire show for final delivery. Make sure to discuss the workflow from edit to audio mix with the editor and audio engineer. Your entire post team should be aware of the process. If you aren't working in the same facility, schedule time to create and deliver audio files.

MUSIC & MEDIA LICENSING

First, as mentioned in the filming section, if you plan on selling or making a profit off it, do not add any music, images or video you do not own. Don't use a popular song that you haven't purchased the rights to. If you don't clear it before hand, these items can be very costly. If you haven't budgeted for them, they can break your budget. If you need music, find someone who is willing to work out a deal with you. Maybe they'll let you use their music to get exposure. When producing shows for networks like MTV and BET I often do licensing deals with music producers. They provide their music with an agreement that once it is placed on air, my team will fill out a cue sheet and list their song credits, so they can receive publishing credit on the back end. A cue sheet is a full log list of songs used within a show. It lists the writer, producer and publishing company. Therefore, every time the show is aired or re-aired they get paid on the backend through publishing.

When producing and directing shows and documentary series, I enjoy using original music. It really brings a flair to my shows because I'm able to select my favorite tracks from the producer's collections. This became a great deal for producers I worked with. They made a nice amount of money on the back end and appreciated having an opportunity to produce music in television and film.

The cue sheet also lists any photos and videos used in the show. When working with networks we usually use sites like Getty and Corbis to acquire images or footage. No one is paying me to mention companies or names, these are just the two I've worked with the most. Feel free to search on-line and find sites that work best for you.

GRAPHICS

If you need graphics, it's best to start creating them before filming starts. When you get to post-production you'll want all the graphic elements finalized. If not finalized, then in the final stages of approval. Sometimes it takes weeks to make graphics and you don't want to extend post days because graphics aren't complete. Plus, if you are working with a network, they need to approve graphic packages in advance. You must schedule time for any executive notes or changes.

If you are handling all these responsibilities by yourself make sure that you show a colleague or professional your edits, so you can get some feedback. If you have a production team, make sure to show the team some of your updated edits. Especially producers, directors, camera operators and writers. Anyone who can add a different perspective. Everyone will appreciate being a part of the post process. Especially if you involve them before the project is released.

Production can be exhilarating and overwhelming at times. Make sure to relax and take it step-by-step. It's a process that must be completed one step at a time. Once you have completed post production, it's time to share your baby with the world!

Chapter 18: PITCHING TO NETWORKS

I FINISHED CREATING MY SHOW. WHAT DO I DO NOW? SHOULD I PITCH MY SHOW TO A NETWORK?

In a moment we will discuss building an audience on social media and YouTube. Before we get to that, let's have a quick conversation about pitching your show to a network.

LET'S TALK ABOUT PITCHING

There is a process to pitching your project to networks and you really need to have a strong package to make it happen.

If you have celebrity talent attached, you might be able to get a network meeting through their talent agent or by pitching your idea to a 3rd party production company. Many of the larger 3rd party production companies have relationships with the networks and will add your show to their show slate of pitches for the season. If your show is a successful pitch then you'll be able to take the next steps which is a negotiation, maybe a contract, a development budget and a pilot.

Here are some of the odds, provided by network executives I work with in the development departments of major networks.

Most development executives listen to an average of 5 pitches per day. That comes out to approximately 25 per week. Broken down into a year, that's 1,250 concepts per year. Some of the show concepts come from independent creative producers who are pitching their first projects. The majority come from agents and production companies

THE PROCESS, THE ODDS REVEALED

Pitches / Day

- They took an average of 5 pitches / day.

- Approximately 25 pitches per week.

- The would add up to 1,250 pitches per year.

- These pitches were a mix of via email and office meeting presentations.

- A small percentage of these pitches came from independent producer/creators.

- Most of the pitches were presented by celebrity agents and 3rd party production companies who have built a reputation with the network and have a successful production record.

- Out of those pitches approximately 4 projects were selected to be presented during weekly development meetings.

- Out of the 1,250 shows presented, approximately 160 shows were pushed up to the next level of development. Nothing was greenlit, it was just moved up to the next executive level for approval.

- The executive development team approved approximately 40 projects for presentation to the Senior Development team.

- Out of the 40 ideas approximately one per month is approved for development.

- That's approximately 10-12 per year.

- Most of these are pilots. It's very rare for a concept to go straight into series.

- Out of the 10-12 projects that are funded for development, only 1-3 series are green lit for air.
- The typical network order is for 6 – 8 episodes.
- Only approximately 1 out of 6 series make it to renewal for Season 2.

This might sound like a tough task but there are plenty of shows that made it to air. Depending on how well your show matches the network, your show could be one of them. Also, if your project wasn't green lit, chances are it was fully developed by the network. At the end of the process your project should be in a great position to pitch to other networks. Therefore, your chances increase that another network will pick it up.

Pitching to a network is like getting a record deal. It sounds great but it's not the only way to get into the industry. There are independent content creators who made a name for themselves plus a huge profit putting out their own content.

Chapter 19: Video Sharing and Monetization

Now that you have created your content (great job!) it's time to share it with the world. There are some options to consider. You can pitch your content to a network with hopes of landing a production deal. That is what we discussed in the previous chapter. Another option is sharing the project you created on a video sharing platform so that you can build a community of loyal fans. Or you can work on finding a distributor or digital aggregator who will represent your project. The digital aggregator is like an agent who will try to sell it to streaming services like Netflix, Amazon or Apple. This option is best if you have a fully completed movie, documentary or series. Even if you are looking to sell to a network or streaming service, building your own digital audience gives you leverage during any meeting or negotiation. If you have a loyal fan base on social media, millions of views on a digital platform and a good product, doors will open for you.

One of the first things they teach you as a child is that sharing is caring. In the world of content that statement is absolutely true. Let's take a moment to discuss the best places to share your video content.

In this digital era, it's essential that you share content with your audience. To be successful you must increase awareness of your project using the digital and social media platforms that are available. It's a great way to communicate with viewers and build your audience. There are so many ways to share. All you have to do is figure out which platforms would work best for you. Video sharing is a great way to showcase your new show and make money. Each platform has a different way for you to monetize your show content.

HMM...MONETIZE YOUR CONTENT? PLEASE EXPLAIN.

Monetizing your content means that you can make money from your shows when sharing them on a digital platform. Once you build up your audience by sharing your show content you'll be able to make money from it. That's one of the themes of this book – turning your idea into money making content!

WHERE CAN I SHARE OR POST MY VIDEO CONTENT? WHICH SITES WILL LET ME MAKE MONEY?

It's time to talk about the top video sharing sites available to content creators. Below, I have listed the top sites for sharing content. There's also a summary of why that platform might work for you and how you can make money.

VIDEO SHARING SITES

Video sharing sites make it easy to find, browse, watch and share content. These sites empower content creators by giving them a direct link to the audience. You're in the driver's seat and whether you're a professional or an amateur because you have a chance to showcase your material. If your content is something people remark about, it can spread like a virus and in turn become 'viral'.

YOUTUBE

YouTube is the most popular video sharing site. Reports say that it brings in nearly 4,950,000,000 views daily. It processes over 3 billion searches per month making it the number 2 search engine behind Google. The platform lets users view, upload and share videos. It's the king of the industry. It's a great place for content creators to build an audience and a digital platform. Look at your channel as your own network. It's the space where you can post your content and build a following. Let people know about your content. Send out links to your content and spread the word. If you need specific stories on how certain YouTube starts became famous and built large following just pick up one of their books for inspiration. One of the things that makes YouTube popular is that it's free. It's free to watch user generated videos and it's free to post them. Another thing that makes YouTube a great platform is that you're able to make money if you build up a large fan base.

HOW CAN I MAKE MONEY ON YOUTUBE?

You can apply for content monetization at any time. It's usually through a form of ad share revenue where you get paid to have advertisers place advertisements on your content. It's like the relationship between television shows and television commercials. You have an audience; your content is getting eye balls and advertisers want to take advantage of those eyeballs.

To qualify, your channel needs 4,000 watch hours and 1,000 subscribers in a 12 months period. Per YouTube "This requirement allows us to properly evaluate new channels and helps protect the creator community."

VIMEO

Another popular video uploading site is Vimeo. It's a U.S. based video sharing website where users can upload, share and view videos. Vimeo allows you to upload videos for free. Reports say that Vimeo has approximately 35 million registered members and the site receives over 240 million viewers per month. On top of that every month viewers watch over 715 million videos. Some of those views can be yours!

HOW DO I MAKE MONEY ON VIMEO?

To make money using Vimeo you must participate in the Video On Demand program. To sell videos you must have

127

a Video PRO account. Vimeo reports that they will notify you every time you make a sale, and you'll get 90% of your sales in a PayPal payment after transaction fees are taken out.

Here's a payment example provided by Vimeo: "For example, a $10 VOD purchase from a buyer based in New York would get you about $7.80. (Taking out transaction fee of $0.48, applicable tax of $0.85 and the revenue sharing of $0.87)".

DAILY MOTION

Dailymotion is one of the biggest video hosting and sharing sites on the internet. It has user-friendly options, allowing millions of users to browse and upload videos by simply searching tags, groups and channels.

To make revenue with Dailymotion you must become a partner.

For more info go to: Dailymotion.com

U STREAM

U Stream is another popular platform for sharing video content. The platform has approximately 100 million active users all over the world. Ustream is owned by IBM and therefore has big-brand support.

HOW DO I MONETIZE MY CONTENT ON U STREAM?

As far as video monetization goes, Pay Per View and other options are available via integration with Cleeng. PPV integration through Cleeng will require an additional fee.

Chapter 20 Social Media Promotion, Spread the Word!

WHERE CAN I BUILD AN AUDIENCE ON SOCIAL MEDIA?

Once you select a video sharing platform for your content you'll have to spread the word and get some viewers. Let your audience know that you have some remarkable content. You can spread the word to your family, friends and fans via social media channels, email, flyers and word of mouth. Feel free to choose whatever will work best, the marketing is up to you. In this digital age you have social media which allows you to communicate directly with your audience.

WHICH ONE SHOULD I USE?

Every social media platform reaches its audience in a different way. Each serves a different purpose. Here's a quick list of the top options that might work for you.

BLOGS

Blogs are an effective way of communicating with your audience. You can create an environment where your audience is expecting content from you. Anyone who signs up for your blog has entered an agreement and has given you permission to communicate with them on a more intimate level.

A content creator can accomplish many things by creating their own blog platform. Some of the benefits are distributing your content to a wide range of viewers at one time, sharing other content that's of interest to you and writing your own thoughts in blog posts.

TWITTER

Twitter is the leading online social networking and micro blogging platform. It allows users to send and read "tweets", with up to 280 characters. Users can provide up to the minute updates about their lives and thoughts. The relationship between a content creator and their audience has never been closer. Whether you update them on daily activities or post links to new content, you get to communicate directly with your followers, in real time.

INSTAGRAM

It's the leading online photo sharing and social networking platform. Users can shoot pictures or short video clips and share them with their online community. If you enjoy taking pictures or sharing photos from your daily life, Instagram is for you. One of the benefits is that it works perfectly with other top social networking sites like Facebook, Twitter, Tumblr and Flickr. Instagram is a great way to keep people updated on memorable moments in your day.

FACEBOOK

YouTube is the King of video sharing and Facebook is the King of social networking. Facebook leads the pack in social engagement and it's a great tool for permission marketing. Anyone who chooses to friend you has agreed to become a part of a content exchange relationship. It's important for a content creator to recognize the power of this relationship. This doesn't mean that you inundate your friends with sales pitches, but it is a place where you can share your work and update your viewers. Another benefit is that you can create a fan page where everyone is more prepared to receive business interactions.

LINKEDIN

LinkedIn is a social networking website for professionals. This is prime time real estate for people in professional fields. A person can choose to use this as a huge digital rolodex of numbers or you can maximize the content sharing opportunities with your professional peers. This platform allows you to establish new relationships while reconnecting with professionals from your past. This is Facebook for professionals. There's more to LinkedIn then sending resumes and searching job contacts. It's a great place to share content and communicate with your supporters from the professional community.

PINTEREST

This is a photo sharing website where users manage photo collections based on their personal interests. Users share pictures and pin other people's photos to their page. It's a fun way to share images. Each snap shot tells a story about you, your likes and your interests.

Here are some final thoughts on Streaming & Sharing:

1. Create content that interests you, be authentic. Make sure there's an audience for it.

2. Set a release schedule so people know when they can expect new content. Alert them to it in advance and build up some interest aka hype!! Set a schedule for your content to be sent out.

3. When you're not sending out original content send out information about you or other projects. Let them know why this information means something to you, engage them. The content you share might be information your audience needs.

4. Share your content with your close circle of supporters then ask them to share it. It's important that you create groups or

tribes of people who will support your creative endeavors. These should be members of your target audience, the people who enjoy your content.

5. If you're a blogger or content site the information you provide can help you establish yourself within the field of interest. Some people even become experts and make a living out of it.

6. Get feedback on your content. Once you send it out people can comment on it and provide important insight that might help improve or further support the content. This builds a relationship between you and the viewer that will benefit both parties.

Don't hold onto that content you produced. Share it! You never know what can happen.

Chapter 21: STREAMING PLATFORMS, V.O.D & DIGITAL AGGREGATORS

<u>I APPRECIATE THE INFORMATION ON VIDEO SHARING SITES AND SOCIAL MEDIA PROMOTION. BUT, HOW DO I SELL MY PROJECT DIRECTLY TO A STREAMING SERVICE LIKE NETFLIX, AMAZON, HULU OR APPLE?</u>

That's a great question. Before we get into how you can sell your project to these platforms, let's talk about what these platforms really are. Each of these services is called a Video On Demand service (VOD). VOD streaming services offer customers a chance to watch shows, films and other content wherever and whenever they want. As the name states, whenever you 'demand' to see your show, it will be available to you. As you know, this exclusive access comes with a fee.

There are three industry standard VOD service options; SVOD, TVOD and AVOD. In your lifetime, you have used one or all of the VOD services. I want to briefly explain the differences between each one. Let's look at how they earn money.

SVOD is an abbreviation for subscription-based video on demand. The service allows you to enter into a subscription agreement where you pay a monthly fee and have access to unlimited programming. Netflix, Amazon Prime and Hulu are examples of an SVOD service.

TVOD is an abbreviation for transactional video on demand service. A TVOD platform normally will not charge anything for you to sign up. It's free to create a user profile but content is not free to watch. Users can only watch content after they purchase it, and they'll pay based on the amount of content they watch. Apple iTunes and Google Play are examples of a TVOD service.

AVOD is an abbreviation for an advertisement supported video on demand service. This model is also free for users. The service makes money by running commercials, brand spots and banners during videos. The idea is that users will watch the free content in return for watching some sponsored advertisements. YouTube is an example of a AVOD platform.

If your plan is to sell your project to a streaming service, you need to work with a digital content aggregator.

WHAT IS AN AGGREAGATOR?

The term 'Aggregator' is used for a company that acts as a gatekeeper between a rights holder and a streaming platform. It's a company that has a pre-established relationship with major buyers in the digital distribution arena and acts as the content gatekeeper.

WHAT DOES THE AGGREGATOR DO?

Aggregators will get your project ready for different VOD streaming platforms. Some of this work includes ingestion, encoding the film, quality control, chapter breaking, closed captioning, administrative expenses and the cost of storing your content in their catalogue.

WHY CAN'T I SELL DIRECTLY TO THE STREAMING SERVICE?

Early on, some of the video streaming companies like iTunes and Hulu would accept submissions from filmmakers on their platform, but that process changed. With an overabundance of films, they became overwhelmed. Now, they only accept digital content from aggregators.

The top video streaming platforms refrain from entering into distribution contracts with individuals or small businesses. Especially companies with one or two films in their catalogue. They want to avoid any administrative issues that can occur when entering into individual contracts with each filmmaker. Their main goal is to provide great content and service to their subscribers. These companies don't want to spend time managing thousands of content creators. The last thing a company like Netflix, Amazon or Apple wants to do is facilitate endless calls and emails from producer and directors. They would prefer to have a reliable third-party company handle those relationships. When it comes to content and delivery, the aggregators have a clearer understanding of what the streaming service requires.

Using a digital aggregator also helps major companies filter content. They are the first line of defense against bad material. If they want to maintain a relationship with the VOD platform they will refrain from pitching low quality films. Instead they will choose to represent films that meet their creative and technical standards. In addition, they will select films that they believe a viewing audience will want to see. It's important for these VOD platforms to stay away from poor quality projects. VOD companies benefit from dealing with companies who have done some of this filtering for them. They also enjoy dealing with companies who have a large catalog of shows, documentaries and films to choose from. Companies like Amazon, Netflix, iTunes and Hulu have built an excellent reputation as platforms with high quality content. I'm sure that they want their reputation to stay that way. Therefore, these platforms refuse to become digital dumping grounds for low quality projects. Creators such as yourself, who spent the time to produce quality content will be rewarded.

HOW MUCH DOES AN AGGREGATOR CHARGE?

The amount an Aggregator charges depends on the company that you deal with. Some companies take between 15-30% of sales. Other companies take an upfront fee plus a yearly maintenance charge. But many of those companies allow the content creator to keep all their profits.

Chapter 22: You Did It!

Thank you for making it to the end of this book. It's been a journey and I truly appreciate your time. We started from an idea, went through pre-production, production, post production and finished with ways to make money from your content. You should be proud of yourself. You are a creative force. It's a great accomplishment, you started something and finished it. This was like going to college. You deserve a degree in Producing and Content Creation. Always acknowledge and appreciate the small victories, they lead to the big victories.

I'm excited to see how you're going to conquer the planet, and I'll be cheering you on along the way. People always talk about having a show idea but very few take steps toward making that idea a reality. I'm proud of you for following your dreams.

In conclusion, I would like to say that there's room for everyone in this creative universe. We must support and encourage each other on this journey. I'm thankful for the people who helped me along the way and I'm thankful to you for reading this book.

Stay positive. Stay blessed. Stay inspired.

Best wishes,

Lee Harris

harriscreativestudio.wordpress.com

Twitter: @LeeMejor | IG: @LeeHarrisOfficial

Made in the USA
Middletown, DE
04 January 2022

57619028R00085